Felipe Matheus Santos e Souza

Diclosulam in different mulches in the suppression of Digitaria

Felipe Matheus Santos e Souza

Diclosulam in different mulches in the suppression of Digitaria

The interference of this herbicide on the germination and growth of crabgrass plants (Digitaria nuda)

ScienciaScripts

Imprint

Cover image: www.ingimage.com

This book is a translation from the original published under ISBN 978-613-0-16258-0.

Publisher:
Sciencia Scripts
is a trademark of
Dodo Books Indian Ocean Ltd. and OmniScriptum S.R.L publishing group

120 High Road, East Finchley, London, N2 9ED, United Kingdom
Str. Armeneasca 28/1, office 1, Chisinau MD-2012, Republic of Moldova, Europe
Managing Directors: Ieva Konstantinova, Victoria Ursu
info@omniscriptum.com

Printed at: see last page
ISBN: 978-620-8-60683-1

ABSTRACT - In a no-till system, mulching the soil is a fundamental part of weed management, as it alters the weed community and also affects the dynamics of herbicides. The aim of this study was to assess the interference with the germination and growth of crabgrass plants (Digitaria nuda Schumach) under different amounts (0, 50 and 100 per cent) and types (black oats (Avena strigosa), ryegrass (Lolium multiflorium), brachiaria grass (Brachiaria ruziziensis), rye (Secale cereale), lupins (Lupinus albus), wheat (Triticum spp.) and triticale (X Triticosecale Wittmack) as mulch for the crops, plus the pre-emergence application of the herbicide diclosulam. The wheat and triticale mulches were more effective in reducing germination and the lupin mulch had no suppressive effect. However, the type of straw had no influence when comparing the dry matter of mattress grass. It was noted that the percentage of mattress grass germination decreased as the amount of mulch increased, except with lupins where the amount had no influence. In all cases, the application of diclosulam was effective, with a reduction in germination and dry matter of mattress grass, and the amount and type of straw did not interfere with the herbicide's action, although in some cases this application may be optional.

Keywords: Soil cover, winter crops, no-till farming

Summary

CHAPTER 1

1. INTRODUCTION

Weeds are a target for control because they can interfere with cultivated plants directly (release of allelopathic compounds and competition for water, light and nutrients) or indirectly (by hosting pests, interfering with the harvest or the quality of the agricultural product, among others), although these plants are often refuges and food sources for predatory insects of important pests of agricultural crops (PITELLI, 1985).

The no-till system currently occupies the position of one of the most important agricultural systems for soil conservation in Brazil (RODRIGUES et al., 2000). Among the crops that adopt this system for their production, soya (Glycine max (L.) Merril) stands out, as it contributed to the implementation of the system in Brazil, initially in the region of Londrina (Paraná) in 1971, on the initiative of the Ministry of Agriculture and farmers from Paraná, and later expanding to various regions of the country (FIDELIS et al., 2003).

Mulch makes a significant contribution to weed control in the no-till system, both through its physical and allelopathic effects (ALMEIDA, 1988). Pitelli and Durigan (2001) also observed that the physical effect of mulch can reduce the chances of survival of weed seedlings with small amounts of reserves in the seeds, which may not be enough to guarantee their survival over the distance travelled through the mulch, until they have access to light and begin the process of photosynthesis. Therefore, among other characteristics, species that have seeds with large amounts of reserves are more likely to perpetuate themselves in this new environment (MICHAELS et al., 1988).

Among the weeds infesting agricultural areas in Brazil are those of the Digitaria genus, which includes around 300 plant species distributed in tropical and subtropical regions of both hemispheres (CANTO-DOROW, 2001a).

Among the weed species known as crabgrass, the following stand out as important in

Brazilian agriculture: Digitaria nuda Schumach, Digitaria ciliaris (Retz.) Koel., Digitaria bicornis (Lam.) Roem. & Schult., Digitaria horizontalis Willd. and Digitaria sanguinalis (L.) Scop. They are plants belonging to the Poaceae family, Panicoidaea subfamily, Paniceae tribe, with a very wide distribution in Brazil, present in the Northeast, Centre-West, Southeast and South regions of the country (DIAS et al., 2005). The D. nuda species is being selected by the repetitive application of the herbicides used to control this weed, characterising a process of population dynamics of specific change in herbicide-tolerant weeds (DIAS et al., 2007).

In the management of agricultural areas targeting weeds, the sequential application of herbicides in pre-planting has led to a delay and/or decrease in weed germination due to the control of the first flush; added to the application of herbicides with a residual effect, there is, as a consequence, a reduction in the use of post-emergence herbicides (COBUCCI; PORTELA, 1999).

The efficiency of herbicides with a residual effect, applied pre-planting, depends on their ability to leach from the straw into the soil, the amount and type of straw, the physical-chemical characteristics of the herbicide molecules and the amount of rainfall after application - these processes define the herbicide's ability to reach the soil (COBUCCI; PORTELA, 2004).

Therefore, studies into the interaction between different types of mulch and herbicides can make weed management increasingly efficient and less toxic to humans and the environment.

The aim of this study was to assess the effect of applying the herbicide diclosulam to different types and amounts of winter crop straw on the control of crabgrass (Digitaria nuda), with a view to simulating no-till farming.

CHAPTER 2

2. LITERATURE REVIEW

2.1. The no-till system

The no-till system (SPD) was introduced in Brazil in the 1970s with the aim of controlling water erosion in the soil and providing greater stability and sustainability in agricultural production compared to the prevailing systems (SOUZA; MADEIRA; FIGUEIREDO, 2014).

Maintaining mulch also improves the maintenance of more water in the soil, reduces spontaneous plants and favours the chemical, physical and biological aspects of the soil, due to the greater contribution of organic matter and nutrients (KURTZ et al., 2013)

In recent decades there has been an increase in the adoption of conservation systems (THEISEN et al., 2000) as well as a major expansion of the no-till system in the Cerrados (KLUTHCOUSKI et al., 2000).

The literature indicates that among the main crops grown in the country, soya stands out as one of the best adapted to the no-till system (KLUTHCOUSKI et al., 2000).

In a no-till system, the remains of dead plants, in addition to improving soil stability, are one of the most important factors that can influence the formation and development of the weed community by reducing and/or altering weed populations, as well as acting on the interception and/or retention of herbicides when applied to the straw, reducing the efficiency of chemical control (RODRIGUES et al., 2000).

2.2. Effects of mulch

Mulch is one of the most important components of the no-till system. The crop residues or mulch present on the soil surface acts as a thermal insulator, as a layer that dissipates the solar radiation that reaches the soil, consequently reducing water evaporation, as well as the soil's hydric and thermal amplitude, and is therefore a key factor for successful agricultural production in the no-till system (PEZZAPONE et al., 1996).

The protection of soils with vegetation cover in temperate regions has been the subject of study by various researchers (DERPSCH; CALERGARI, 1992). However, in tropical regions, where the climate favours the rapid decomposition of crop residues and the quality and durability of plant residues must be taken into account (ALVES; COGO; LEVIEN, 1995), these studies are relatively new.

Mulch can act on weeds in three different ways: physically, chemically and biologically (PITELLI; PITELLI, 2004). As for the physical effect, mulch can alter the quality and quantity of incident light, affecting the germination process of some species such as positive photoblasts (they need a certain wavelength to germinate), as well as reducing the germination of seeds that need a wide temperature range to start the germination process, or even reducing the chances of survival of species with insufficient reserves in the diaspores to overcome the mulch and start the photosynthetic process.

In terms of biological aspects, mulch provides the conditions for the installation of a dense and varied microbiocenosis in the topsoil that can use weed seeds and seedlings as energy sources and can also create a safe haven for some natural enemies.

In terms of chemical effects, mulch can release chemical substances into the environment that can interact with weeds, inhibiting or stimulating their growth and development (allelopathy) (MONQUERO et al. 2009).

2.3. **The genus** Digitaria

On the American continent, Brazil stands out as the country with the largest number of Digitaria species, 26 of which are native, nine of which are exclusive, and 12 of which are exotic. The specific richness in the different regions of Brazil is more or less the same, with a significant decrease in the northern region, thus characterising the species' preference for inhabiting open formations such as natural fields and cerrados (DIAS et al., 2007). Some species have been used as fodder, while others stand out for being crop weeds (CANTO-DOROW; LONGHI-WAGNER, 2001).

Digitaria nuda occurs in tropical regions on all continents, but is apparently most abundant in Africa (VELDKAMP, 1973), where it was originally described. It had already been mentioned for Rio Grande do Sul by Cavalheiro and Barreto (1981), and is considered adventitious in the state. This species occurs at low altitudes and is sterile, apparently preferring sandy soils (CAVALHEIRO; BARRETO, 1981; VELDKAMP, 1973).

Digitaria nuda has two spikelets per node, 2-4 mm long, 0.6-0.7 mm wide, lanceolate; lower glume absent; upper glume 0.9-1.2 mm long, 0.3-0.5 mm wide, up to 1/2 the length of the lower lemma, acute, hairy, acute trichomes, whitish, sparse, on the margins and between the veins, surpassing the apex of the glume; lower lemma 2-2.1 mm long, 0.6-0.7 mm wide, acute, hairy on the margins and between the lateral veins, acute trichomes, dense, exceeding the apex of the lemma by 0.2-0.4 mm; lower palea absent; upper anthecium 1.8-2.2 mm long, 0.4-0.6 mm wide, with a bisexual flower, lanceolate, acute, with papillae in longitudinal rows, light brown at maturity. In terms of the size of the spikelets, it is similar to D. horizontalis, from which it differs in having a lower glume and a lower lemma with trichomes that do not extend beyond its apex (CANTO-DOROW, 2001b).

2.5. Behaviour of herbicides

Research shows that mulches reduce the density of weed populations. However, many of these studies show that even with the reduction in population density, some control measures are still necessary, of which herbicides stand out (FORNAROLLI et al., 1998).

Cobucci et al. (2004) stated that the efficiency of herbicides with a residual effect applied pre-planting depends on their ability to leach from the straw into the soil, the amount and type of straw, the physical-chemical characteristics of the herbicide molecules and the amount of rainfall after application - these processes define the herbicide's ability to reach the soil. This fact has attracted the attention of several researchers. Studies carried out by Pastana (1972) in the USA showed that when atrazine was applied to soil with oat mulch, the weight of the dry matter of the weeds that grew there was greater than in soil without mulch, suggesting that part of the product was retained in the straw. The reduction in weed dry matter weight in mulched soil was achieved by increasing the dose of the product.

Bauman (1977) found that when atrazine was applied to rye stubble, 30 per cent less of the product reached the soil than when it was sprayed directly onto bare soil.

The origin and quantity of the mulch can influence the ability of a residual herbicide to reach the soil in the no-till system. Lowder and Weber (1979) mention that the retention of atrazine by maize and oat plant remains may explain, in part, the need to use higher doses of the herbicide on these mulches than in conventional tillage; they also mention that more atrazine was removed by rain from oat plant remains than from maize remains. Banks and Robinson (1982) found that, as the amount of wheat mulch increases, there is a corresponding reduction in the amount of metribuzin that reaches the soil in the no-till soya system.

Other factors cited by some authors that have a major influence on the retention of herbicides by straw are the amount and timing of rainfall after application. Banks and Robinson (1986) found an increase in the amount of metolachlor, alachlor and acetochlor on the soil surface under wheat mulch as the amount of irrigation water increased. Lowder and Weber (1979) found that when 100 mm of water is applied immediately after the application of atrazine, 87% of the herbicide is removed from the mulch, but if it is applied on the 7th day, 77% of the product is removed. Some authors, however, point out that water solubility is the main characteristic that gives a herbicide greater or lesser ability to reach the soil in the no-till system. Comparing the herbicides metolachlor and alachlor, Strek and Weber (1982) found that the former was less intercepted by wheat mulch, possibly because it is more soluble than the latter; being water soluble, when the first rain falls after application, they are washed from the straw into the soil (ALMEIDA, 1992).

Another characteristic that can exert this influence is volatility. Clomazone, for example, being relatively volatile (RODRIGUES; ALMEIDA 2011), could be even more volatile when applied on mulch than on bare soil (THELAN et al.,1986), reducing the amount of the product that reaches the soil in the no-till system by more than 40 per cent (MILLS et al.,1989).

2.6. The herbicide diclosulam

For the herbicide diclosulam, Rodrigues and Almeida (2011) classified it as a herbicide from the triazolo pyrimidine sulphonanilides chemical group, indicated for the control of dicotyledons in incorporated pre-sowing or pre-emergence in the soya crop. This compound acts by inhibiting the enzyme acetolactate synthase (ALS), which is essential for the synthesis of the amino acids valine, leucine and isoleucine. Diclosulam and its metabolites have very low partition values (Kd), which demonstrates their potential for vertical movement, facilitated by their solubility in water (124 ppm at pH 7.0 and 117 ppm at pH 5.0), acid ionisation constant (pKa) of 4.09 and octanol/water coefficient (Kow) of 1.42 (YODER et al., 2000).

Currently, there are not many studies on the herbicide diclosulam in the control of monocots. Timossi (2005) observed that the herbicide diclosulam was able to leach through sugar cane straw, acting efficiently in the control of weeds of the genera *Ipomoea* and *Merremia,* showing evidence of medium solubility and low retention in the straw. Timossi (2005) also observed that the residual of the herbicide diclosulam was important for weed control and did not cause problems for soya as a successor crop.

Recently, Minozzi et al. (2012) observed that the addition of diclosulam to the pre-sowing desiccation of weeds promoted greater control and reduced the emergence of several species during the soya cycle, especially Commelina benghalensis L. and Alternanthera tenella Colla, and that a single application of glyphosate in post-emergence was satisfactory.

CHAPTER 3

3. MATERIAL AND METHODS

The experiment was carried out from 29/10/2013 to 17/12/2013 at the Department of Biology Applied to Agriculture, Faculty of Agricultural and Veterinary Sciences, UNESP, Jaboticabal Campus, SP.

The region has a tropical climate classified as Cwa (KOPPEN, 1948), with an average annual rainfall over the last 38 years of 1,443.8 mm, characterised by a distribution of rainfall concentrated in the summer and a dry winter.

Six species of winter crops were used in the experiment. black oats (Avena strigosa Schreb.), ryegrass (Lolium multiflorium L.), rye (Secale cereale L.), lupins (Lupinus albus L.), wheat (Triticum spp.) and triticala (X Triticosecale Wittmack).), wheat (Triticum spp.) and triticale (X Triticosecale Wittmack); as well as a tropical forage plant, brachiaria grass (Brachiaria ruziziensis R. Germ. & C. M. Evrard) as mulch.

The species used as mulch were sown on 26/03/2013 in the experimental area of the PET Agronomy group at the Faculty of Agricultural and Veterinary Sciences, UNESP, Jaboticabal Campus, SP, with a location of 21° 14' 14.34" S and 48° 17' 26.38" O and an elevation of 598 metres from sea level.

Sixty days before sowing the species used as mulch, twenty soil samples were collected from the experimental area The samples were homogenised to form a single sample and the chemical properties of the sample were analysed according to the procedures for soil analysis in Brazil (EMBRAPA, 1997), shown in Table 1.

Table 1. Chemical properties of the soil used to plant the species used as mulch in 2013. Jaboticabal, SP, 2014.

PH	**M.O.**	**P Resin**	**K**	**Ca**	**Mg**	**H+Al**	**SB**	**T**	**V**
in $CaCl_2$	g/dm^3	mg/dm^2	---------------------mmolc/dm^3 ------------------						%
5,3	19	34	3,3	24	15	25	42,3	67,3	63

During the crop and forage cycle, no pesticides or inputs were applied, except sowing fertiliser in the amount of 300 kg ha^{-1} of NPK formulation 2-20-20.

The plant materials from the mulches were collected on 10/10/2013. To determine the amount of mulch, a circumference the size of the diameter of the pots (23 cm) was used and the plants inside the circumference were cut off close to the ground, characterising the sample unit of 100% of the amount of mulch. The sample unit with 50% of the amount of mulch had a similar collection, however, only half of the plants inside the circumference were cut. In the treatment without mulch, the soil was left completely bare, i.e. 0% straw. The reproductive parts, such as inflorescences and pods, were removed, simulating the natural deposition of vegetative material after harvest.

Once collected, the samples were placed in paper bags and transported to the greenhouse, where they were left to dry for 19 days.

Each experimental unit consisted of a plastic pot with a capacity for five litres of soil and a mixture of clay soil and sand was used as the substrate for the pots, in a 3:1 v/v ratio, respectively.

On 29/10/2013, the dry matter of the seven species was placed on the substrate, thus characterising the mulch. The following day, mattress grass was sown in the soil beneath the mulch, with an average population of 190.25 seeds per pot.

The experimental design used was entirely randomised, with the treatments arranged in a 7x3x2 factorial scheme, in four replications, in which the factors were made up of the seven mulches, in three quantities, with and without herbicide application.

The diclosulam herbicide was applied on 31/10/2013 at 16:00 h, in a controlled environment, at a rate of 30 g/ha of commercial product, i.e. 25.21 g a.i./ha. The herbicide was applied using a constant pressure (CO_2) knapsack sprayer, fitted with a boom with four XR11002 nozzles, spaced every 0.5 m, and adjusted for a spray volume of 200L ha^{-1}, applied over the plants and at a distance of 1m. The temperature at the time of application was 28°C. The newly applied pots were left in a protected environment for a day and then returned to the open environment. The first rainfall after herbicide application was 25 mm, one day after application (DAA).

When the pots were taken out of the protected environment, a plastic tarpaulin was stretched over them to prevent the influence of weeds from the outside environment, as well as a screen under the pots to prevent dry matter from being dispersed by the wind. Once all the material had stabilised in the pots, approximately one week later, the screen and plastic sheeting were removed.

At 48 DAA, the emerged plants were counted to determine the germination percentage of each experimental unit. Also at 48 DAA, the mattress grass weeds that had germinated in each pot were collected for dry matter determination. They were cut close to the ground, placed in paper bags and dried in an oven with forced air circulation at 50°C for 96 hours. Once dry, they were weighed to determine the dry matter of the mattress grass weeds.

The data was normalised using the Box-Cox transformation (OSBORNE, 2010). They were then analysed in a factorial scheme with three factors (7x3x2), with the type of mulch (black oats, ryegrass, brachiaria grass, rye, lupins, wheat and triticale), amount of mulch (0, 50 and 100%) and herbicide application (present or absent) and the means were compared using the Tukey test at 5%. The normality of the residuals was checked using the Shapiro-Wilk test (SHAPIRO; WILK, 1965). The statistical analyses were carried out in the R software (R DEVELOPMENT CORE TEAM, 2014) using the agricolae , asbio , ExpDex & MASS packages. Analysis was also carried out with two factors (3x2), amount of mulch (0, 50 and 100%) and herbicide application (present or absent) and the means were also compared using the Tukey test at 5%.

CHAPTER 4

4. RESULTS AND DISCUSSION

4.1. Germination and dry matter as a function of mulch type, amount of mulch and herbicide application

The germination percentage and dry matter of crabgrass in relation to the type of mulch (black oats, ryegrass, brachiaria grass, rye, lupins, wheat and triticale) amount of mulch (0, 50 and 100%) and herbicide application (present or absent) showed no significant interaction, even after normalising the data (Table 2).

Table 2. Summary of the analysis of variance for germination percentage and dry matter of cocksfoot grass, as a function of treatment. Jaboticabal, SP, 2014.

Sources of variation	GL[1]	Mean squares	
		Germination (%)[2]	Dry mass (g)[3]
Type of mulch (A)	6	3,12 *	2.19 NS
Amount of mulch (B)	2	68,50 **	3,91 *
Herbicide application (C)	1	50,80 **	144,57 **
A x B	12	1.23 NS	1.24 NS
A x C	6	0.95 NS	1.29 NS
B x C	2	15,64 **	1.45 NS
A x B x C	12	0.91 NS	1.21 NS
Waste	125		
CV (%)		35,63	24,86

[1] GL: degree of freedom; Transformed data: [2] - sqrt (%G + 0.5), [3] - sqrt (MS + 0.5); NS: not significant; * significant at 5%; ** significant at 1%, by the F test; CV: coefficient of variation.

4.1.1. Germination of crabgrass depending on the amount of mulch and herbicide application

For the germination percentage of crabgrass, there was a significant interaction between the amount of mulch and herbicide application, after normalising the data (Table 2).

In relation to the treatments with herbicide application (Table 3), there was a significant difference in the germination percentage of the treatment without mulch compared only to the treatment with the highest amount of mulch (100%). As for the treatments without herbicide application, there was a significant reduction in the germination percentage in all the mulch treatments (50 and 100%) when compared to the treatment without mulch (0%).

Table 3. Germination percentage of crabgrass at 48 days after application (DAA) as a function of the amount of cover, with and without the herbicide diclosulam. Jaboticabal, SP, 2014.

Variables analysed	100%	50%	0%	F
WITH HERBICIDE	2,334 Ba	3.376 Ab	4,082 Ab	10,50 **
NO HERBICIDE	2,810 Ca	4,311 Ba	7.398 Aa	73,24 **
F	1.51 NS	5,94 *	74,61 **	

Averages followed by distinct uppercase letters in the row and distinct lowercase letters in the column differ by the Tukey test (p<0.05). Transformed data: - sqrt (%G + 0.5); NS: not significant; * significant at 5% and ** significant at 1%, by the F test.

As for the amount of mulch (Table 3), only with 100% was there no significant difference between the presence or absence of herbicide application. The treatments with 0 and 50% mulch differed significantly, with germination being higher in the treatments where no herbicide was applied, showing that when the mulch is not very dense, despite reducing the weed population, some control measures are still necessary, often using herbicides (FORNAROLLI et al., 1998).

4.1.2. Germination of crabgrass depending on the type of cover

When analysing the types of plant material, there was a statistical difference between the treatments (Table 2).

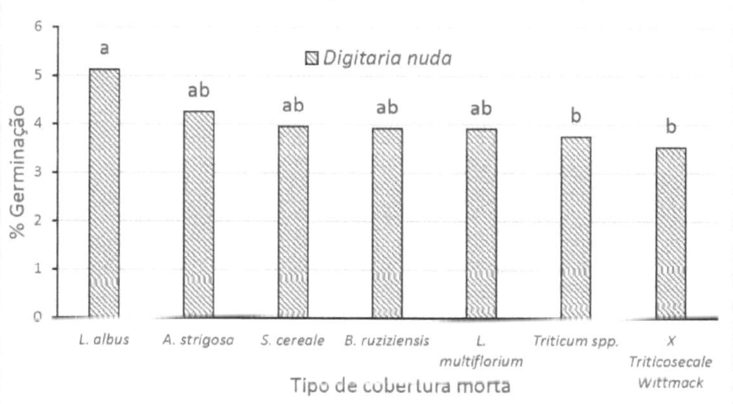

Averages followed by different letters differ by the Tukey test (p<0.05) and averages followed by equal letters do not differ by the Tukey test (p>0.05).

Figure 1. Germination percentage of crabgrass at 48 days after application as a function of mulch types. Jaboticabal, SP, 2014.

When comparing the different types of mulch (Figure 1), it was observed that the presence of lupins resulted in the highest germination percentage, differing statistically only from wheat and triticale, while the other species did not differ. Lamego et al. (2013) found different results when they analysed the emergence of blackgrass (Conyza bonariensis) in winter crop mulches, where the wheat treatments showed an increase in the number of weeds compared to other mulches such as black oats and ryegrass.

4.1.3. Dry matter of cocksfoot grass in all types of cover

The average dry matter values of the cocksfoot grass did not show any interaction between the type of mulch and the amount of mulch or the application of herbicide, nor between the amount of mulch and the application of herbicide, even after normalising the data (Table 2), so the amount of mulch (Figure 2) and whether or not chemical control was carried out with the herbicide were analysed separately (Figure 3). A comparison was also made between the types of mulch (Figure 4).

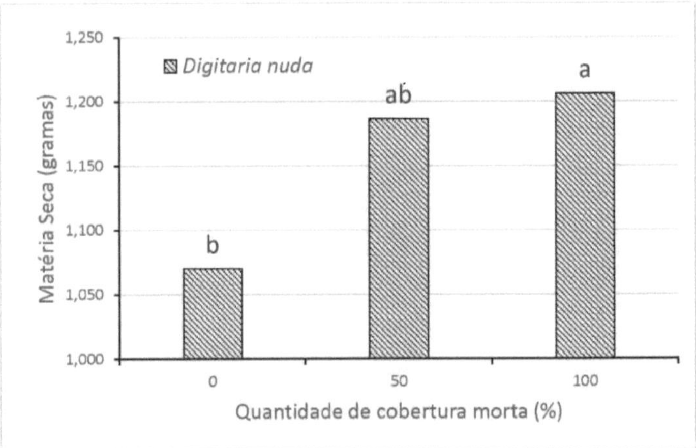

Averages followed by different letters differ by the Tukey test ($p<0.05$) and averages followed by equal letters do not differ by the Tukey test ($p>0.05$).

Figure 2. Dry matter of cocksfoot grass at 48 days after application as a function of the amount of mulch. Jaboticabal, SP, 2014.

The amount of mattress grass dry matter (Figure 3) increased as the amount of mulch increased, with the treatment with no mulch being statistically lower than the treatment with the highest amount of mulch (100 per cent).

There was a significant difference in the dry mass of the cocksfoot grass in the treatments with and without the application of the herbicide diclosulam (Figure 4), with the dry mass being higher in the treatments without the herbicide. As for the dry mass of the mattress grass in relation to the type of mulch (Figure 4), there was no statistical difference and the species did not influence the dry mass of the weed.

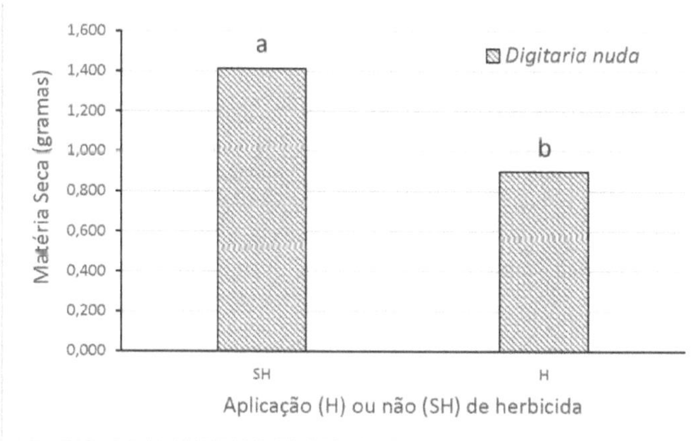

Averages followed by different letters differed according to the Tukey test (p<0.05).

Figure 3. Dry matter of cocksfoot grass at 48 days after application as a function of the herbicide diclosulam. Jaboticabal, SP, 2014.

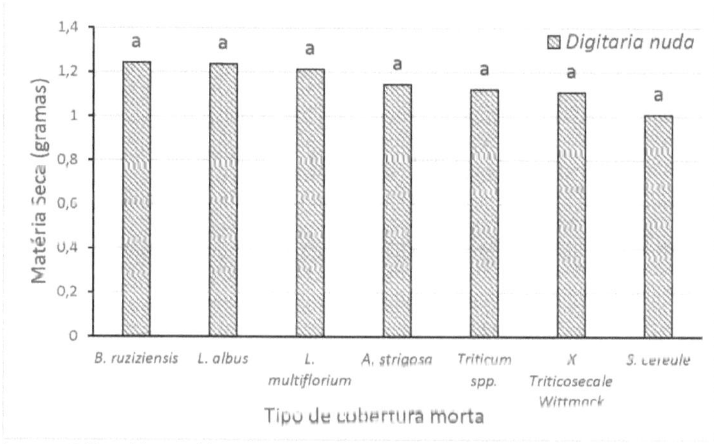

Averages followed by equal letters do not differ according to Tukey's test (p>0.05).

Figure 4. Dry matter of cocksfoot grass at 48 days after application as a function of mulch types. Jaboticabal, SP, 2014.

4.2. Black oats

4.2.1. Germination of crabgrass in black oat cover crops

The results of mattress grass germination on black oat cover had no significant interaction, even after normalising the data, and a simple effect analysis was carried out, comparing the effect of the amount of straw and the effect of the herbicide (Table 4).

Table 4 Summary of the analysis of variance for the percentage of mattress grass germination in response to the amount of black oat mulch, in the presence and absence of the herbicide diclosulam. Jaboticabal, SP, 2014.

Sources of variation	GL[1]	Mean squares
		Germination (%)
Amount of straw (A)	2	12,01 **
Herbicide (B)	1	21,27 **
A x B	2	3.57 NS
Waste	17	
CV (%)		50,32

[1] GL: degree of freedom; NS: not significant; ** significant at 1% by the F test; CV: coefficient of variation.

The treatments with black oat mulch (Figure 5), regardless of the amount, had a lower germination rate of cushion grass, which differed statistically from the treatment without mulch, confirming the reports of Kissmann (1991) who observed that black oats are an excellent forage, providing good soil cover and also reducing the pressure of various invasive species. The reduction in germination in the treatments with 50 and 100 per cent mulch may be related not only to the physical effects but possibly to the sum of these effects and the allelopathic action of the black oat straw (ALMEIDA, 1988; RABELO, 1991; HAGEMANN, 2010).

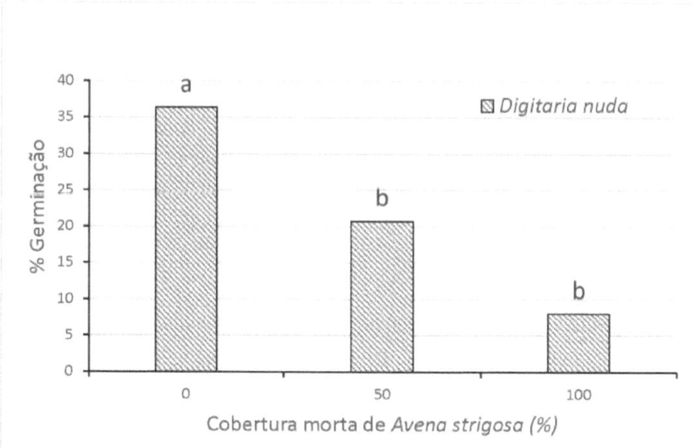

Averages followed by different letters differ according to Tukey's test ($p<0.05$) and averages followed by equal letters do not differ according to Tukey's test ($p>0.05$).

Figura 5. Germination percentage of crabgrass at 48 days after application as a function of amounts of black oat mulch. Jaboticabal, SP, 2014.

The germination percentage in relation to herbicide application (Figure 6) in the treatment without diclosulam herbicide application was higher and statistically different when compared to the treatment with application, the latter being more effective in eventual control.The germination of mattress grass with 100% black oat straw was similar to germination with herbicide application, thus indicating that the use of this species as mulch would dispense with or reduce the use of herbicide (RODRIGUES et al. 2000).

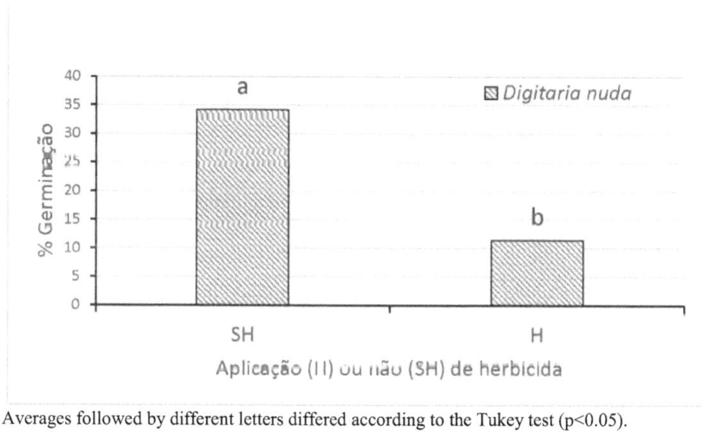

Averages followed by different letters differed according to the Tukey test ($p<0.05$).

Figure 6. Germination percentage of crabgrass at 48 days after application of the herbicide diclosulam on black oat mulch. Jaboticabal, SP, 2014.

4.2.2. Dry matter of cocksfoot grass covered with black oats

As for the dry matter of cocksfoot grass in black oat cover in relation to the amount of cover and herbicide application, there was no significant interaction, even after normalising the data. Therefore, a simple effect analysis was carried out, comparing the effect of the amount of straw and the effect of the herbicide separately (Table 5).

Table 5 Summary of the analysis of variance for the dry mass of cocksfoot grass in response to the amount of black oat mulch, in the presence and absence of the herbicide diclosulam. Jaboticabal, SP, 2014.

Sources of variation	GL[1]	Mean squares Dry mass (g)[2]
Amount of straw (A)	2	0.69 NS
Herbicide (B)	1	37,76 **
A x B	2	0.56 NS
Waste	17	
CV (%)		41,10

[1] GL: degree of freedom; Transformed data: [2] - log (MS + 1); NS: not significant; ** significant at 1% by the F test; CV: coefficient of variation.

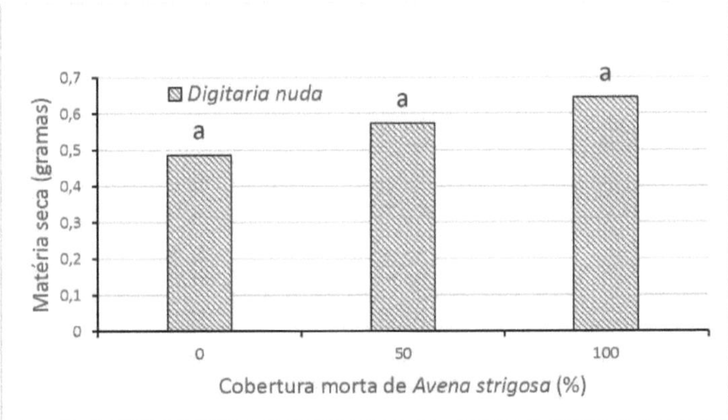

Transformed data: - log (MS + 1); Averages followed by equal letters do not differ by Tukey's test (p>0.05).

Figure 7. Dry matter of cocksfoot grass at 48 days after application as a function of the amount of black oat mulch. Jaboticabal, SP, 2014.

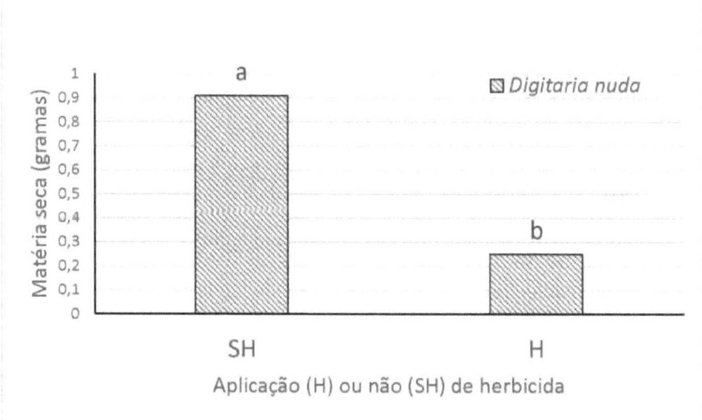

Transformed data: - log (MS + 1); Averages followed by different letters differ by Tukey's test ($p<0.05$).

Figure 8. Dry matter of cocksfoot grass at 48 days after application of the herbicide diclosulam on black oat mulch. Jaboticabal, SP, 2014.

Regarding the amount of black oat mulch (Figure 7), there was no significant difference in the dry matter of cushion grass, regardless of the amount of straw.

However, dry matter (Figure 8) in the treatment without diclosulam differed statistically when compared to the treatment with diclosulam.

4.3. Ryegrass

4.3.1. Germination of cocksfoot grass in ryegrass cover

The results of mattress grass germination in ryegrass mulch, like black oat mulch, had no significant interaction, even after normalising the data. Therefore, a simple effect analysis was carried out, comparing the effect of the amount of straw and the effect of the herbicide separately (Table 6).

Table 6. Summary of the analysis of variance for germination percentage (%G) of cocksfoot grass in mulch treatments (ryegrass), in three mulch quantities (0, 50 and 100%), in the presence and absence of herbicide application. Jaboticabal, SP, 2014.

Sources of variation	GL[1]	Mean squares
		Germination (%)[2]
Amount of straw (A)	2	12,22 **
Herbicide (B)	1	9,10 *
A x B	2	1.38 NS
Waste	18	
CV (%)		33,18

[1] GL: degree of freedom; Transformed data: [2] - log (%G + 1); NS: not significant; * significant at

5% and ** significant at 1%, by the F test; CV: coefficient of variation.

Mattress grass germination (Figure 9) in the treatment without mulch differed significantly only from the treatment with the highest amount of ryegrass mulch (100%). Moraes et al. (2011) observed something similar, where it was found that ryegrass straw provided greater germination of mattress grass, with the exception of the highest level of straw, where there was a reduction in emergence.

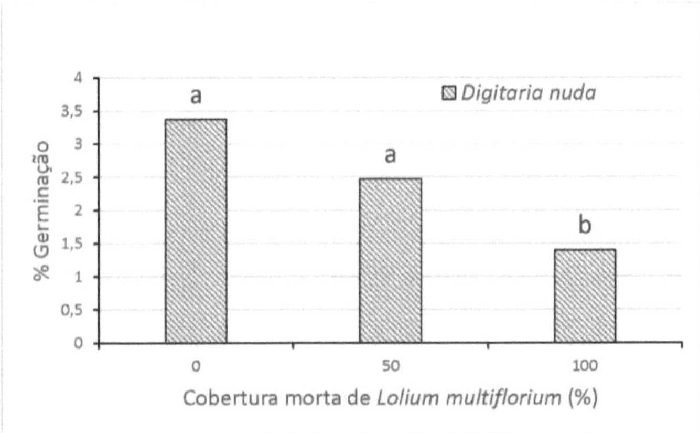

Transformed data: - log (%G + 1); Means followed by different letters differ by Tukey's test (p<0.05) and means followed by equal letters do not differ by Tukey's test (p>0.05).

Figure 9. Average values obtained for germination percentage of crabgrass at 48 days after application (DAA) in the mulch treatments (0, 50 and 100%). Jaboticabal, SP, 2014.

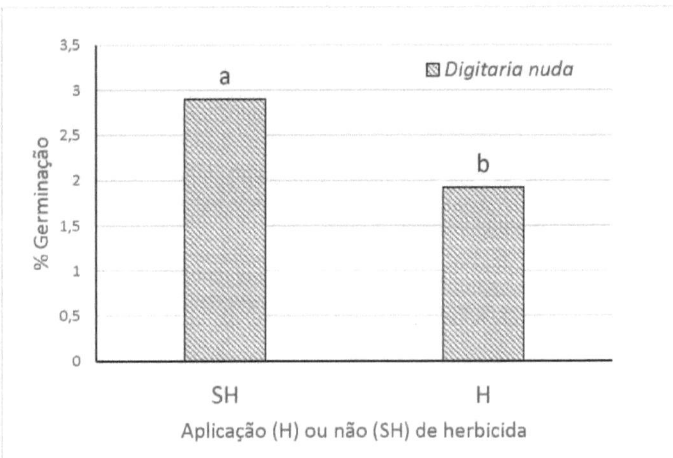

Transformed data: - log (%G + 1); Averages followed by different letters differ by Tukey's test ($p<0.05$).

Figure 10. Average values obtained for germination percentage of crabgrass at 48 days after application (DAA) in treatments with ryegrass mulch where there was herbicide application (H) or no application (SH). Jaboticabal, SP, 2014.

The germination percentage in the treatment without application of the herbicide diclosulam was higher (Figure 10), and was statistically different when compared to the treatment with application, indicating that the association between straw and herbicide may represent more effective weed control.

4.3.2. Dry matter of cocksfoot grass in ryegrass cover

Table 7. Summary of the analysis of variance for dry matter (g) of cocksfoot grass in mulch treatments (ryegrass), in three mulch quantities (0, 50 and 100%), in the presence and absence of herbicide application. Jaboticabal, SP, 2014.

Sources of variation	GL[1]	Mean squares
		Dry mass (g)[2]
Amount of straw (A)	2	5,04 *
Herbicide (B)	1	41,45 **
A x B	2	4,41 *
Waste	18	
CV (%)		53,70

[1] GL: degree of freedom; Transformed data: [2] - log (MS + 1); * significant at 5% and ** significant at 1%, by the F test; CV: coefficient of variation.

On the other hand, the results for dry matter of cushion grass under ryegrass cover showed a significant interaction between the amount of cover and herbicide application (Table 7).

Table 8. Average values obtained for dry matter (g) of crabgrass at 48 days after application (DAA) in treatments with ryegrass mulch in three mulch quantities (0, 50 and 100%), with and without herbicide application. Jaboticabal, SP, 2014.

Variables analysed	**100%**	**50%**	**0%**	**F**
WITH HERBICIDE	0.407 Ab	0.340 Ab	0.245 Ab	0.07 NS
NO HERBICIDE	1,670 Ba	2,940 Aa	1,142 Ba	9,37 **
F	8,75 *	37,10 **	5,03 *	

Transformed data: - log (MS + 1); Averages followed by distinct uppercase letters in the row and distinct lowercase letters in the column differ by Tukey's test ($p<0.05$). NS: not significant; * significant at 5% and ** significant at 1%, by the F test.

The treatments with herbicide application showed no significant difference in terms of mattress grass dry matter (Table 8), and the amount of straw had no influence on the results. However, in the treatments without herbicide application, the intermediate amount of straw (50%) had a positive influence on the development of cushion grass, differing statistically from the other treatments, while without mulch and the maximum amount of straw (100%) had reduced dry matter. Moraes et al. (2011) observed that ryegrass was the cover crop that, on the soil surface or incorporated, showed the greatest reduction in the dry matter of Digitaria spp. when spread in large quantities.

4.4. Brachiaria grass

The results with brachiaria cover differed from those of the cover plants mentioned above, with an interaction between the amount of cover and herbicide application, both for germination and dry matter of cushion grass (Table 9).

Table 9. Summary of the analysis of variance for germination percentage and dry matter (g) of cocksfoot grass in mulch treatments (brachiaria), in three mulch quantities (0, 50 and 100%), in the presence and absence of herbicide application. Jaboticabal, SP, 2014.

Sources of variation	GL[1]	Mean squares	
		Germination (%)	Dry mass (g)[2]
Amount of straw (A)	2	13,42 **	3.29 NS
Herbicide (B)	1	5,53 *	64,20 **
A x B	2	6,69 *	6,51 *
Waste	18		
CV (%)		62,04	37,89

[1] GL: degree of freedom; Transformed data: [2] - log (MS + 1); NS: not significant; * significant at 5% and ** significant at 1%, by the F test; CV: coefficient of variation.

3.5.1. Germination of cocksfoot grass in brachiaria grass cover

In relation to the treatments with herbicide application (Table 10), it was observed that the presence or absence of brachiaria cover had no influence on weed germination.

On the other hand, when the herbicide was not applied (Table 10), the presence of straw was fundamental in attenuating the germination of crabgrass, and in the treatments without mulch, germination was statistically higher than in the others. **Table 10.** Average values obtained for the percentage of cocksfoot grass germination at 48 days after application (DAA) in the treatments with brachiaria grass mulch in three mulch quantities (0, 50 and 100%), with and without herbicide application. Jaboticabal, SP, 2014.

Variables analysed	100%	50%	0%	F
WITH HERBICIDE	2,235 Aa	21.287 Aa	18,000 Ab	2.77 NS
NO HERBICIDE	7,357 Ba	14.717 Ba	54.665 Aa	17,33 **

F	0.35 NS	0.58 NS	17,98 **

Averages followed by distinct uppercase letters in the row and distinct lowercase letters in the column differ by the Tukey test ($p<0.05$). NS: not significant; ** significant at 1% by the F test.

Ribeiro et al. (2010) observed in their research areas that when the growth of the species B. ruziziensis was sufficient to cover the soil, forming a good amount of vegetation cover, the reduction in the number of weeds was very significant from the outset.

Regarding the amount of mulch (Table 10), it was observed that for the mulch treatments, regardless of the amount (50 or 100 per cent), the presence or absence of herbicide did not interfere with germination. Numerous studies have shown that the physical effect of mulch can be enhanced by the residue of some species that release substances with allelopathic properties, greatly increasing the control action of weeds (PITELLI, 1987) and thus reducing the need for chemical control.

The treatment without mulch (Table 10) showed a significant difference in terms of reduced germination when the herbicide was applied, thus demonstrating the importance of applying diclosulam in conventional systems.

4.4.2. Dry matter of cocksfoot grass on brachiaria grass cover

There was no significant difference in the dry matter of cocksfoot grass in the amounts of brachiaria mulch compared to the treatments with herbicide application (Table 11).

Table 11. Average values obtained for dry matter (g) of cocksfoot grass at 48 days after application (DAA) present in the treatments with brachiaria grass mulch in three mulch quantities (0, 50 and 100%), with and without herbicide application. Jaboticabal, SP, 2014.

Variables analysed	100%	50%	0%	F
WITH HERBICIDE	0.124 Ab	0.424 Ab	0.212 Ab	1.49 NS
NO HERBICIDE	1.477 Aa	1.002 Ba	0.761 Ba	8,31 *
F	57,31 **	10,48 *	9,43 *	

Transformed data: - log (MS + 1); Averages followed by distinct uppercase letters in the row and distinct lowercase letters in the column differ by Tukey's test ($p<0.05$). NS: not significant; * significant at 5% and ** significant at 1%, by the F test.

However, in the treatments without the herbicide diclosulam (Table 11), the treatment with the maximum amount of brachiaria straw (100%) differed from the other treatments with higher dry mass values . The treatments with 50% mulch and no mulch did not differ from each other.

Analysing each amount of mulch in relation to the presence or absence of herbicide (Table 11) showed that all the treatments differed statistically, with the dry mass being higher in the treatments without herbicide application for the three amounts of mulch (0, 50 and 100%).

The latter results show an inversion of values when compared to the germination percentage, where while dry matter increases with an increase in the amount of mulch, germination decreases with the same ratio. An increase in the proportion of weeds favours an increase in competitive pressure. In general, when plants are found in high densities, the value of each individual as a competitive element is diminished, as is its growth potential, since the development of the population is controlled by the resource that is least available in the environment (PITELLI, 1987).

4.5. Rye

4.5.1. Germination of crabgrass on rye cover

For the germination of cocksfoot grass in rye cover in relation to the amount of cover and the application of herbicides, there was a significant interaction, as can be seen in Table 12.

Table 12. Summary of the analysis of variance for the percentage of mattress grass germination, present in the mulch treatments (rye), in three amounts of mulch (0, 50 and 100%), in the presence and absence of herbicide application. Jaboticabal, SP, 2014.

Sources of variation	GL[1]	Mean squares
		Germination (%)
Amount of straw (A)	2	13,24 **
Herbicide (B)	1	7,17 *
A x B	2	6,43 *
Waste	18	
CV (%)		58,45

[1] GL: degree of freedom; NS: not significant; * significant at 5% and ** significant at 1%, by the F test; CV: coefficient of variation.

There was no significant difference in the germination of crabgrass in the amounts of rye mulch compared to the treatments with herbicide application (Table 13).

However, when the herbicide was sprayed (Table 13), it was possible to see a statistical difference between the treatments, with germination in the treatment with no coverage being higher than in the other treatments (50 and 100 per cent coverage).

When the amount of mulch was analysed in relation to the application of the herbicide (Table 13), germination was similar for the mulch treatments (50 and 100%), and only in the treatment without mulch was there a statistical difference, with reduced germination in the treatment with the application of diclosulam.

Table 13. Average values obtained for germination percentage of cocksfoot grass at 48 days after application (DAA) in treatments with rye mulch in three mulch quantities (0, 50 and 100%), with and without herbicide application. Jaboticabal, SP, 2014.

Variables analysed	100%	50%	0%	F
WITH HERBICIDE	5.782 Aa	16,820 Aa	18,000 Ab	7.17 NS
NO HERBICIDE	8,412 Ba	15,637 Ba	54.665 Aa	18,33 **
F	0.10 NS	0.02 NS	19,80 **	

Averages followed by distinct uppercase letters in the row and distinct lowercase letters in the column differ according to Tukey's test ($p < 0.05$).

3.5.2. Dry matter of cushion grass on rye cover

As for the dry matter of cocksfoot grass in rye mulch in relation to the amount of mulch and herbicide application, there was no significant interaction, even after normalising the data. Therefore, a simple effect analysis was carried out, comparing the effect of the amount of straw and the effect of the herbicide separately (Table 14). Regarding the amount of rye mulch (Figure 11), no significant difference could be observed in the dry matter of mattress grass, regardless of the amount of straw.

Table 14. Summary of the analysis of variance for dry matter (g) of cocksfoot grass in mulch treatments (rye), in three mulch quantities (0, 50 and 100%), in the presence and absence of herbicide application. Jaboticabal, SP, 2014.

Sources of variation	GL[1]	Mean squares
		Dry mass (g)
Amount of straw (A)	2	0.30 NS
Herbicide (B)	1	14,39 **
A x B	2	0.04 NS
Waste	18	
CV (%)		88,12

[1] GL: degree of freedom; NS: not significant, * significant at 5% and ** significant at 1%, by the F test; CV: coefficient of variation.

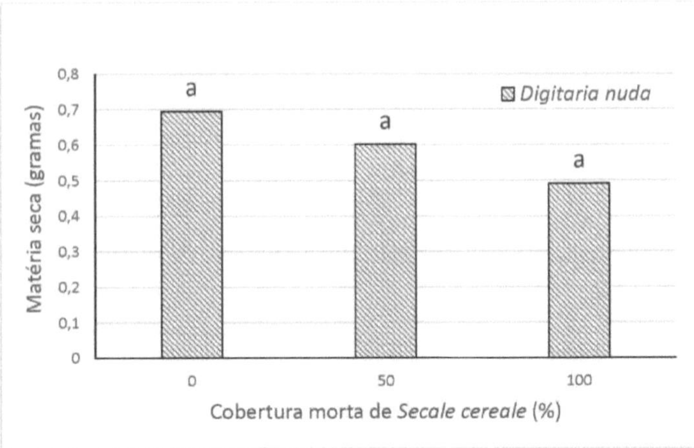

Averages followed by equal letters do not differ according to Tukey's test (p>0.05).

Figure 11. Average values obtained for dry matter (g) of cocksfoot grass at 48 days after application (DAA) present in the treatments with rye mulch in the amounts of mulch (0, 50 and 100%). Jaboticabal, SP, 2014.

Regarding the presence or absence of herbicide application (Figure 12), the dry matter in the treatment without diclosulam herbicide application was higher and statistically different when compared to the treatment with diclosulam application.

Balbinot Jr. et al. (2007) found that, in a no-till system, a consortium of ryegrass, black oats and rye can help to reduce weed infestation in the subsequent crop; however, in many situations there is still a need for appropriate intervention with herbicides to avoid a reduction in productivity due to interference from these weeds.

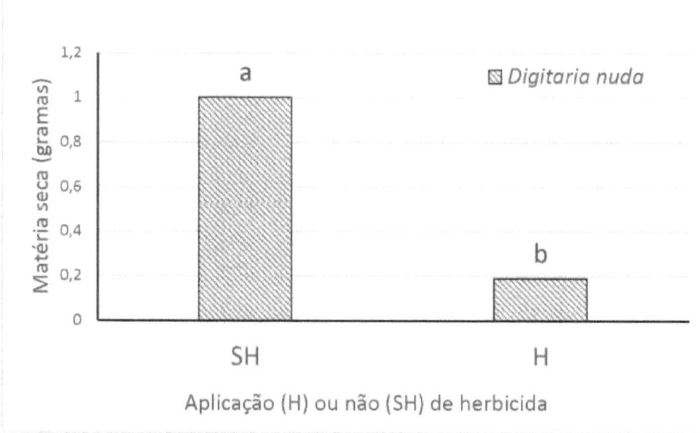

Averages followed by different letters differed according to Tukey's test ($p<0.05$).

Figure 12. Average values obtained for dry matter (g) of cocksfoot grass at 48 days after application (DAA) present in treatments with rye mulch where there was herbicide application (H) or no application (SH).

4.6. Lupins

4.6.1. Germination of cocksfoot grass in lupin mulch

There was a significant interaction between the amount of mulch and herbicide application in relation to the germination of crabgrass in lupin cover, as can be seen in Table 15.

Table 15. Summary of the analysis of variance for the percentage of mattress grass germination, present in the mulch treatments (lupins), in three amounts of mulch (0, 50 and 100%), in the presence and absence of herbicide application. Jaboticabal, SP, 2014.

Sources of variation	GL[1]	Mean squares
		Germination (%)
Amount of straw (A)	2	1.95 NS
Herbicide (B)	1	3,99 *
A x B	2	4,56 *
Waste	18	
CV (%)		50,10

[1] GL: degree of freedom; NS: not significant; * significant at 5% by the F test; CV: coefficient of variation.

In the treatments with herbicide application (Table 16), there was no significant difference in the germination of crabgrass, regardless of the presence or absence of mulch from lupins. This result may be related to the retention of the herbicide in the straw, although according to Carbonari et al. (2008), the herbicide diclosulam has a good leaching capacity in straw from small amounts of rain.

When no herbicide was sprayed (Table 16), the presence of lupin debris, regardless of the amount (50 and 100 per cent) influenced the reduction in germination, with the values of the treatments with cover being lower and statistically different from those of the treatment without cover (0 per cent).

For germination within each amount of mulch in relation to the presence or absence of herbicide (Table 16), it was only possible to see a statistical difference in the treatment without mulch, where the germination percentage was higher when no herbicide was applied.

Table 16. Average values obtained for germination percentage of crabgrass at 48 days after application (DAA) in the treatments with lupin mulch in three mulch quantities (0, 50 and 100%), with and without herbicide application. Jaboticabal, SP, 2014.

Variables analysed	100%	50%	0%	F
WITH HERBICIDE	23,257 Aa	27,070 Aa	18,000 Ab	0.40 NS
NO HERBICIDE	21,550 Ba	27,202 Ba	54.665 Aa	6,10 *
F	0.03 NS	0.00 NS	12,07 *	

Averages followed by distinct uppercase letters in the row and distinct lowercase letters in the column differ by the Tukey test ($p<0.05$). NS: not significant; * significant at 5% by the F test.

4.6.2. Dry matter of cocksfoot grass under lupin cover

Table 17. Summary of the analysis of variance for dry matter (g) of cocksfoot grass in mulch treatments (lupins), in three mulch quantities (0, 50 and 100%), in the presence and absence of herbicide application. Jaboticabal, SP, 2014.

Sources of variation	GL[1]	Mean squares
		Dry mass (g)[2]
Amount of straw (A)	2	0,78 *
Herbicide (B)	1	0,77 *
A x B	2	0.45 NS
Waste	18	
CV (%)		20,13

[1] GL: degree of freedom; Transformed data: [2] - MS ^A 0.169653; NS: not significant; * significant at 5% by the F test; CV: coefficient of variation.

The dry matter of cocksfoot grass in lupin cover in relation to the amount of cover and herbicide application showed no significant interaction, even after normalising the data. Therefore, a simple effect analysis was carried out, comparing the effect of the amount of straw and the effect of the herbicide separately (Table 17).

The amount of lupin mulch (Figure 13) significantly affected the dry matter of the mattress grass. The treatments with lupin mulch (50 and 100 per cent) provided greater biomass accumulation, which differed from the treatment without mulch (0 per cent).

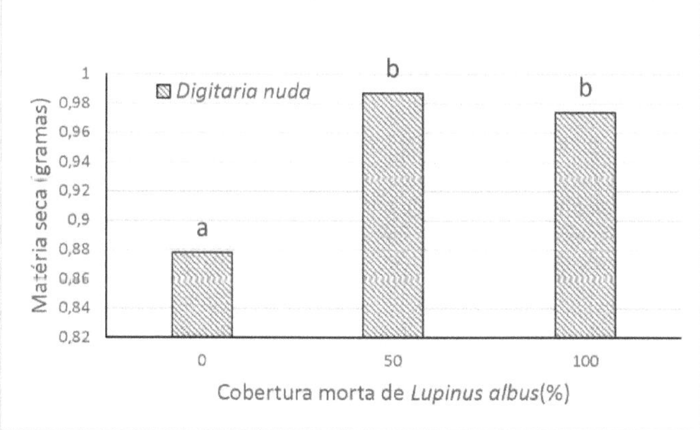

Transformed data: - MS ^A 0.169653; Means followed by different letters differ by Tukey's test ($p<0.05$) and means followed by equal letters do not differ by Tukey's test ($p>0.05$).

Figure 13. Average values obtained for dry matter (g) of crabgrass at 48 days after application (DAA) present in the treatments with lupin mulch in the amounts of mulch (0, 50 and 100%). Jaboticabal, SP, 2014.

The environment with the presence of lupin mulch can be favourable for the development of weeds, just as it was for the bean plant, as observed by Derpsch and Calegari (1992) where the highest bean yields were obtained when the bean was sown directly on white lupin crop remains.

In relation to the presence or absence of the herbicide diclosulam (Figure 14) there was a reduction in dry matter in the treatment with the presence of the application, and this treatment differed statistically from the treatment without the application. As a result, the benefits of mulching can be utilised without the damage caused by the presence of weeds.

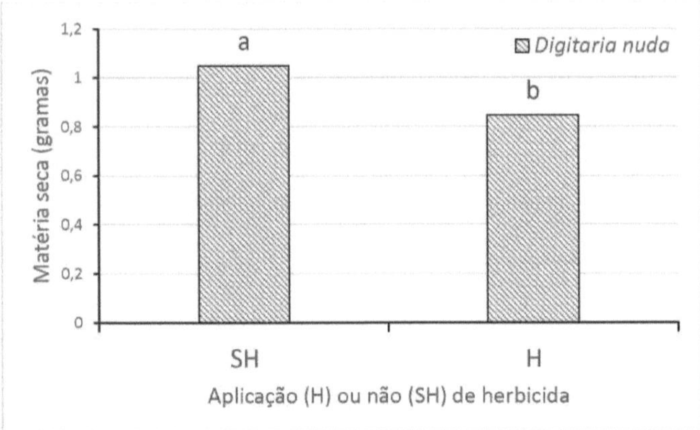

Transformed data: - MS ^ 0.169653; Means followed by different letters differ according to Tukey's test ($p<0.05$).

Figure 14. Average values obtained for dry matter (g) of crabgrass at 48 days after application (DAA) present in the treatments with lupin mulch where there was herbicide application (H) or no application (SH).

4.7. Wheat

4.7.1. Germination of cocksfoot grass on wheat cover

The germination of crabgrass in wheat cover in relation to the amount of cover and herbicide application showed a significant interaction, as can be seen in Table 18.

There was no significant difference in the germination of cocksfoot grass in the amounts of wheat mulch compared to the treatments with herbicide application (Table 19). The treatments without herbicide showed a significant difference, with the treatment without mulch (0%) differing from the other treatments (50 and 100%), which did not differ from each other. As for germination of the cocksfoot grass in the amounts of cover as a function of the presence of herbicide, there was only one difference

statistically significant when the treatments were in the absence of straw (0%), with the absence of herbicide providing the highest germination percentage.

Table 18. Summary of the analysis of variance for the percentage of mattress grass germination, present in the mulch treatments (wheat), in three amounts of mulch (0, 50 and 100%), in the presence and absence of herbicide application. Jaboticabal, SP, 2014.

Sources of variation	GL[1]	Mean squares
		Germination (%)
Amount of straw (A)	2	20,89 **
Herbicide (B)	1	5,32 *
A x B	2	11,09 **
Waste	18	
CV (%)		55,86

[1] GL: degree of freedom; NS: not significant; * significant at 5% and ** significant at 1%, by the F test; CV: coefficient of variation.

Table 19. Average values obtained for germination percentage of cocksfoot grass at 48 days after application (DAA) in treatments with wheat mulch in three quantities (0, 50 and 100%), with and without herbicide application. Jaboticabal, SP, 2014.

Variables analysed	100%	50%	0%	F
WITH HERBICIDE	8,147 Aa	13,405 Aa	18,000 Ab	0.97 NS
WITHOUT HERBICIDE	3,942 Ba	9,327 Ba	54.665 Aa	31,01 **
F	0.35 NS	0.33 NS	26,85 **	

Averages followed by distinct uppercase letters in the row and distinct lowercase letters in the column differ by the Tukey test ($p<0.05$). NS: not significant; * significant at 5% and ** significant at 1%, by the F test.

3.5.3. Dry matter of cocksfoot grass in wheat cover

As for the dry matter of cocksfoot grass in wheat mulch in relation to the amount of mulch and the application of herbicides, no significant interaction could be verified. Therefore, a simple effect analysis was carried out, comparing the effect of the amount of straw and the effect of the herbicide separately (Table 20).

The dry matter of the cocksfoot grass (Figure 15) was not affected by the different amounts (0, 50 and 100 per cent) of wheat mulch, i.e. there was no significant difference regardless of the amount of wheat straw.

Table 20. Summary of the analysis of variance for dry matter (g) of cocksfoot grass in mulch treatments (wheat), in three mulch quantities (0, 50 and 100%), in the presence and absence of herbicide application. Jaboticabal, SP, 2014.

Sources of variation	GL1	Mean squares
		Dry mass (g)
Amount of straw (A)	2	1.49 NS
Herbicide (B)	1	18,64 **
A x B	2	1.05 NS
Waste	18	

CV (%)	69,82

[1] GL: degree of freedom; NS: not significant; * significant at 5% and ** significant at 1%, by the F test; CV: coefficient of variation.

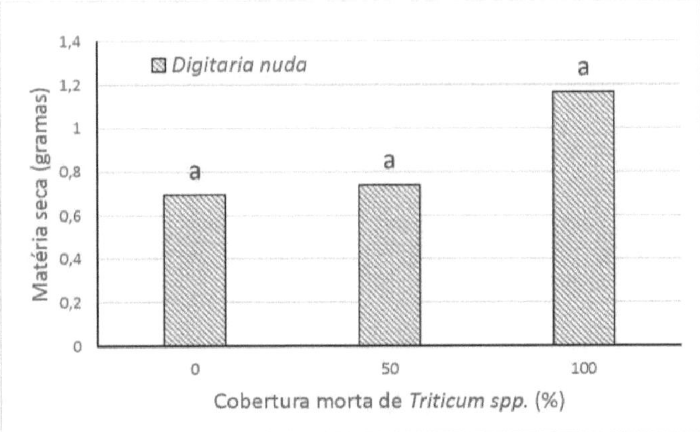

Averages followed by equal letters do not differ according to Tukey's test (p>0.05).

Figure 15. Average values obtained for dry matter (g) of cocksfoot grass at 48 days after application (DAA) present in the treatments with wheat mulch in the amounts of mulch (0, 50 and 100%). Jaboticabal, SP, 2014.

However, the presence of straw on the soil tends to favour the reduction of germination of many species, but in some cases straw can alter the germination of some weed species; this may occur, according to Correia et al. (2006), due to a reduction in the soil's daily temperature range, moisture conservation or even chemical, physical and/or biological improvements. With regard to dry matter as a function of the presence or absence of herbicide application, the treatment without application of the herbicide diclosulam was higher, being statistically different when compared to the treatment with application (Figure 16).

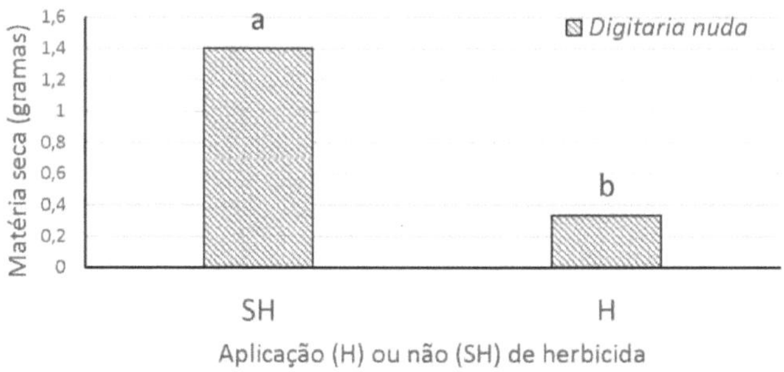

Averages followed by different letters differed according to Tukey's test ($p<0.05$).

Figure 16. Average values obtained for dry matter (g) of cocksfoot grass at 48 days after application (DAA) present in the treatments with wheat mulch where there was an application of herbicide (H) or no application (SH).

4.8. Triticale

4.8.1. Germination of cushion grass in triticale mulch

The germination of crabgrass in triticale cover in relation to the amount of cover and herbicide application showed a significant interaction, as can be seen in Table 21.

As for the germination of crabgrass in the amounts of wheat mulch in relation to the treatments with herbicide application, no significant difference was found (Table 22).

Table 21. Summary of the analysis of variance for the percentage of mattress grass germination, present in the mulch treatments (triticale), in three amounts of mulch (0, 50 and 100%), in the presence and absence of herbicide application. Jaboticabal, SP, 2014.

Sources of variation	GL [1]	Mean squares
		Germination (%)
Amount of straw (A)	2	30,31 **
Herbicide (B)	1	24,73 **
A x B	2	7,15 *
Waste	18	
CV (%)		52,16

[1] GL: degree of freedom; NS: not significant; * significant at 5% and ** significant at 1%, by F test; CV: coefficient of variation.

However, in the treatments without herbicide application, there was a statistical difference with a 75.24 and 84.14 per cent reduction in the percentage of mattress grass germination in the treatments with 50 and 100 per cent mulch, respectively, compared to the treatment without mulch (0 per cent).

With regard to the amount of this cover, we observed similar germination in each quantity, with the application of the herbicide being optional, except in the treatment without triticale cover, where the application of the herbicide diclosulam significantly reduced weed germination (67.07%).

Table 22. Average values obtained for the percentage of crabgrass germination at 48 days after application (DAA) present in the treatments with triticale mulch in the maximum (100), average (50) and no mulch (0) amounts, also where there was herbicide application (H) or application (SH).

Variables analysed	100%	50%	0%	F
WITH HERBICIDE	2,497 Aa	3,155 Aa	18,000 Ab	4.03 NS
NO HERBICIDE	8,672 Ba	13,537 Ba	54.665 Aa	33,43 **
F	0.99 NS	2.82 NS	35,20 **	

Averages followed by distinct uppercase letters in the row and distinct lowercase letters in the column differ by the Tukey test ($p<0.05$). NS: not significant; ** significant at 1% by the F test.

4.8.2. Dry matter of cocksfoot grass under triticale cover

The results for dry matter of cocksfoot grass in wheat cover in relation to the amount of cover and herbicide application showed no significant interaction. Therefore, a simple effect analysis was carried out, comparing the effect of the amount of straw and the effect of the herbicide separately (Table 23).

Table 23. Summary of the analysis of variance for the dry matter of cocksfoot grass present in the mulch treatments (triticale), in three amounts of mulch (0, 50 and 100%), in the presence and absence of herbicide application. Jaboticabal, SP, 2014.

Sources of variation	GL[1]	Mean squares
		Dry mass (g)
Amount of straw (A)	2	0.34 NS
Herbicide (B)	1	47,57 **
A x B	2	0.41 NS
Waste	18	
CV (%)		47,02

[1] GL: degree of freedom; NS: not significant; * significant at 5% and ** significant at 1%, by the F test; CV: coefficient of variation.

There was no significant difference in the dry matter of cushion grass, regardless of the amount of triticale straw (Figure 17).

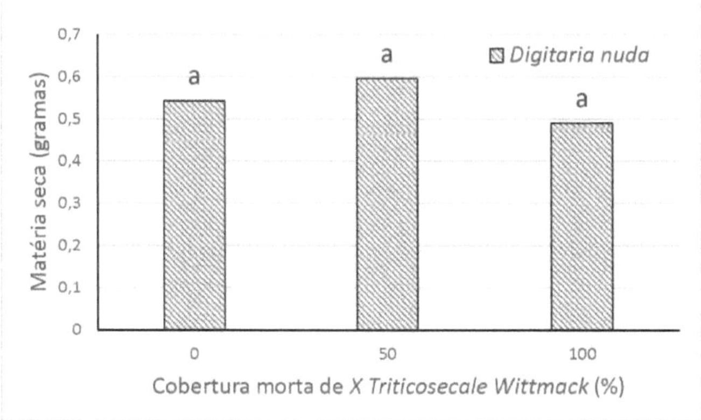

Averages followed by equal letters do not differ according to Tukey's test ($p>0.05$).

Figura 17. Average values obtained for dry matter (g) of crabgrass at 48 days after application (DAA) present in the treatments with triticale mulch in the mulch quantities (0, 50 and 100%). Jaboticabal, SP, 2014.

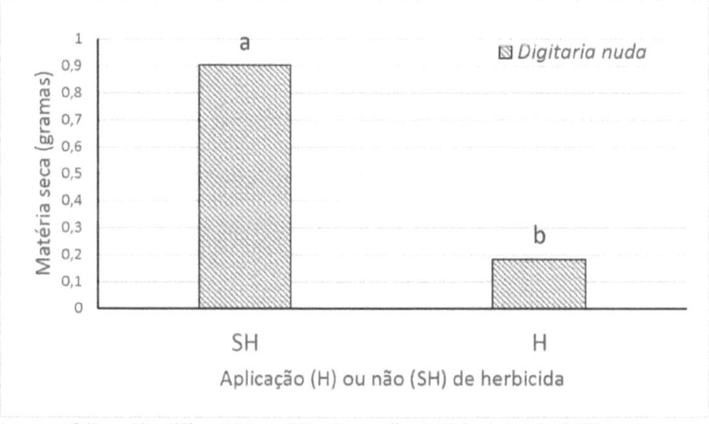

Averages followed by different letters differed according to Tukey's test ($p<0.05$).

Figure 18. Average values obtained for dry matter (g) of crabgrass at 48 days after application (DAA) present in the treatments with triticale mulch where there was herbicide application (H) or no application (SH).

In relation to the presence or absence of herbicide application (Figure 18), dry matter differed statistically, and dry matter was higher in the treatment without application.

CHAPTER 5

CONCLUSIONS

It can be concluded that the application of diclosulam was effective, even though this herbicide is more suitable for dicotyledons, regardless of the type of mulch, reducing the germination and development, including flowering, of mattress grass.

Lupin cover alone was not effective in suppressing the germination and development of the weed, but when combined with the herbicide it did reduce it. Wheat and triticale mulches reduced the germination of crabgrass when compared to lupins. However, the type of straw did not influence the amount of dry matter of the crabgrass.

REFERENCES

ALMEIDA, F. S. **Allelopathy and plants**. Londrina: IAPAR, 1988. 60 p. (Technical Circular, 53).

ALMEIDA, F.S. Residual herbicides in different soil preparation systems. **Pesquisa Agropecuária Brasileira**, Brasília, v.27, n.4, p.596-601, 1992.

ALMEIDA, F.S.; RODRIGUES, B.N. **Guia de herbicidas**. 2. ed. Londrina: Ed. dos Autores, 1988. 603 p.

ALVES, A.G.C.; COGO, N.P.; LEVIEN, R. Relations of soil erosion with the persistence of mulch. **Revista Brasilieira de Ciencia do Solo**, Viçosa, MG, v.19, p. 127-132, 1995.

BALBINOT JR., A.A.; MORAES, A.; BACKES, R.L. Effect of winter cover crops and their timing on weed infestation in maize. **Planta daninha** vol.25 no.3 Viçosa July/September. 2007

BANKS, P.A.; ROBINSON, E.L. Soil reception and activity of acetochlor, alachlor and metolachlor as affected by wheat (Triticumaestivum) straw and irrigation. **Weed Science**, Champaign, v.34, p.607-611, 1986.

BANKS, P.A.; ROBINSON, E.L. The influence of straw mulch on the soil reception and persistence of metribuzin. **Weed Science**, Champaign, v.30, p.164-168,1982.

BAUMAN, T.T. Amount and persistence of atrazine in soil with three tillage systems. **Dissertation Abstracts International**, London, v .37, n .10, p .4794 - 4795, 1977.

CANTO-DOROW, T. S. Digitaria Heister ex Haller. In: WANDERLEY, M. G. L.; SHEPHERD, G. J.; GIULIETTI, A. M. (Ed.). **Phanerogamic flora of the state of São Paulo**. São Paulo: HUCITEC, 2001a. p. 143-150.

CANTO-DOROW, T. S. **The genus** Digitaria **Haller (Poaceae - Panicoideae - Poniceae) in Brazil**. 2001b. 386 f. Thesis (Doctorate in Plant Science) - Federal University of Rio Grande do Sul, Porto

Alegre, 2001.

CANTO-DOROW, T. S.; LONGHI-WAGNER, H. M. Taxonomic novelties in Digitaria Haller (Poaceae) and new citations for the genus in Brazil. **INSULA**, Florianópolis, v. 30, p. 21-34, 2001.

CARBONARI, C.A.; MESCHEDE, D.K.; CORREA, M.R.; VELINI, E.D.; TOFOLI, G.R. Efficacy of the herbicide diclosulam in association with sorghum straw in the control of Ipomoea grandifolia and Sida rhombifolia. **Planta daninha**, Viçosa, v.26, n.3, p.657- 664 ,2008

CAVALHEIRO, E. M.; BARRETO, I. L. The indigenous or spontaneous species of the genus Digitaria, Heister ex Haller (Gramíneae) occurring in Rio Grande do Sul.

Anuário Técnico do Inst.ituto de Pesquisas Zootécnica "Francisco Osório", Porto Alegre, v. 8, p. 171-315, 1981.

COBUCCI, T.; PORTELA, C.M.O.; SILVA, W.; NETO MONTEIRO, A. Behaviour of herbicides with residual effect in different cover crops of bean. **Planta Daninha**, Viçosa, MG, v.22, n 4, oct./dec. 2004

COBUCCI, T; PORTELA, C. M. O. Sequential application of herbicides applied in pre-planting to control weeds in the bean crop. In: REUNIÃO NACIONAL DE PESQUISA DE FEIJÃO, 1999, Salvador. **Expanded abstracts...** Santo Antônio de Goiás: Embrapa Arroz e Feijão, 1999. p. 465-468

CORREIA, N. M. et al. Influence of the type and quantity of plant residues on weed emergence. **Planta Daninha**, v. 24, n. 2, p. 245-253, 2006.

DERPSCH, R.; CALEGARI, A. **Plants for winter green manure**. Londrina: Iapar, 1992. 80p. (Circular 73)

DIAS, A.C.R.; CARVALHO, S.J.P.; NICOLAI, M.; CHRISTOFFOLETI, P.J. Problematic occurrence of different species of crabgrass (Digitaria spp.) in sugarcane crops. **Planta Daninha**, Viçosa, MG, v.25, n.3, p. 489-499, 2007.

DIAS, N.M.P.; CHRISTOFFOLETI, P.J.; TORNISIELO, V.L. Taxonomic identification of crabgrass species infesting sugarcane crops in the state of São Paulo and efficacy of herbicides in controlling Digitaria nuda. **Bragantia**, Campinas v. 64, n. 3, p. 389-396, 2005.

FORNAROLLI, D.A.; RODRIGUES, B.N.; LIMA, J.; VALÉRIO, M.A. Influence of mulch on the behaviour of the herbicide atrazine. **Planta Daninha**, Viçosa, v. 16, n. 2, p. 97-107, 1998.

HAGEMANN, T.R. Allelopathic potential of oat leaf aqueous extracts on ryegrass and peanut. **Bragantia**, Campinas, v. 69, n. 3, p. 509-518, 2010.

KISSMANN, K. G. Avena strigosa Schreb. In: **Weeds and Noxious Plants**. BASF.1991, São Paulo. 1ª ed. 603 p.

KOPPEN, W. **Climatologia**: con un estudio de los climas de la tierra. Mexico: Fondo de Cultura Económica, 1948. 479 p.

KLUTHCOUSKI. J.; FANCELLI, A.L.; DOURADO-NETO, D.; RIBEIRO, C.M.; FERRARO, L.A. Soil management and the yield of soya, maize, beans and rice in no-till farming. **Scientia**

Agricola, Piracicaba, v. 57, n. 1, p. 97-104, 2000.

KURTZ, C.; ERNANI, P.R.; PAULETTI, V.; MENEZES JUNIOR, F.O.G.; VIEIRA NETO, J. Onion productivity and conservation affected by nitrogen fertilisation in the no-till system. **Horticultura Brasileira**, Brasília, v.31, n.4, p.559-567, 2013.

LAMEGO, F.P.; KASPARY, T.E.; RUCHEL, Q.; GALLON, M.; BASSO, C.J.; SANTI, A.L. Management of glyphosate-resistant Conyza bonariensis: winter covers and pre-sowing herbicides for soya. **Planta daninha**, Viçosa, v.31, n.2, p.433- 442, 2013.

LAVORENTI, A. et al. Behaviour of diclosulam in samples of a dystrophic red latosol under no-till and conventional tillage. **Revista Brasileira de Ciencia do Solo**, Viçosa, MG, v. 27, n. 2, p. 183-190, 2003.

LOWDER, S.W.; WEBER, J.B. Atrazine retention by crop residues in reduced tillage systems. In: SOUTHERN WEED SCIENCE SOCIETY, 32, 1979, Georgia. **Proceedings**...p .303-307.

MICHAELS, H.J.; BENNER, B.; HARTGERINK, A.P.; LEE, T.D.; RICE, S.; WILLSON, M.F.; BERTIN, R.I. Seed size variation: magnitude, distribution, and ecological correlates. **Evolutionary Ecology**, London, v.2, n.2, p.157-166, 1988.

MILLS, J.A.; WITT, W.W.; BARRETT, J. Effects of tillage on the efficacy and persistence of clomazone in soybean (Glycine max). **Weed Science**, Champaign, v.37, p.217-222, 1989.

MINOZZI,G.B.; PEREIRA,P.A.; MONQUERO, P.A. Diclosulam in the management of weeds in the RR soya crop. In: CONGRESSO BRASILEIRO DA CIÊNCIA DAS PLANTAS DANINHAS, 28, 2012, Campo Grande. **Proceedings**...

MONQUERO, P.A.; AMARAL, L.R.; INÁCIO, E.M.; BRUNHARA, J.P.; BINHA, D.P.; SILVA, P.V.; SILVA, A.C. Effect of green manures on the suppression of weed species. **Planta Daninha**, Viçosa, MG, v.27, n.1, p. 85-95, 2009.

MORAES, P.V.D.; AGOSTINETTO, D.; PANOZZO, L.E.; TIRONI, S.P.; GALON, L.; SANTOS, L.S. Allelopathy of cover plants on the surface or incorporated into the soil in the control of Digitaria spp. **Planta daninha**, Viçosa, v.29, n.especial, p.963-973, 2011

OSBORNE, J.W. Improving your data transformations: Applying the Box-Cox transformation. **Practical Assessment, Research & Evaluation**, v.15, n.12, 9p. 2010.

PASTANA, F.I. Effect of herbicide retention by soil mulch on weed control and maize production with minimum cultivation. **Bragantia**, Campinas, v.31, p.259 - 274, 1972.

PEZZAPANE, J.E.M.; CUNHA, G.M.; ARNSHOLZ, E.; JÚNIOR, M.C. Soil temperature as a function of mulching with coffee straw. **Revista Brasileira de Agrometeorologia**, v.4, p.7-10,1996.

PITELLI, R. A.; DURIGAN, J. C. Weed ecology in the no-till system. In: ROSSELO, R. D. No-till **farming in the Southern Cone**. Montevideo: Procisur, 2001. v.1, p. 203-210

PITELLI, R.; PITELLI, R.L.. M. Biology and ecophysiology of weeds. In: VARGAS, L.; ROMAN, E.S. **Manual de manejo e controle de plantas daninhas.** Bento Gonçalves: Embrapa Uva e Vinho, 2004. Chap 2, p. 29-56.

PITELLI, R. A. Weed competition and control in agricultural areas. **Série Técnica IPEF**, Piracicaba, v.4, n.12, p.1-24, 1987.

PITELLI, R. A. Interference of weeds in agricultural crops. **Informação Agropecuária**, Fortaleza, v.11, n.129, p.16-27, 1985.

R DEVELOPMENT CORE TEAM. **R**: a language and environment for statistical computing. Vienna: R Foundation for Statistical Computing, 2011. Available at:< http://www.R-project.org/>. Accessed on: 10 Mar. 2014

RABELO, R. R. Survey of the current situation of crop rotation and green manure in the Centre-South region. In: South-Central Meeting on Green Fertilisation and Crop Rotation, 3. Cascavel, PR, 1991. **Minutes of the II Centre-South Meeting on Green Fertilisation and Crop Rotation.** Cascavel, OCEPAR, 1991. p. 53-95.

RIBEIRO, C.A; MATEUS, R.P.G; FORNAROLLI, D.A; MATEUS, R.P.G; DEBASTIANI, R.; NOEDI, B.N.; GAZZIERO, D.L.P. Effect of the presence of Brachiaria ruziziensis in consortium with maize (Zea mays) on weed suppression. XXVII Brazilian Congress of Weed Science. Ribeirão Preto, 2010. **Proceedings**...

RODRIGUES, B.N.; ALMEIDA, F. S. **Guia de herbicidas**. 6. ed. Londrina: Grafmarke, 2011. 697 p.

RODRIGUES, B.N.; LIMA, J.; YADA, I.F.U.; ULBRICH, A.V.; FORNAROLLI, D.A. Influence of mulch on the retention of *imazaquin* in no-till soya. **Planta Daninha**, Viçosa, MG, v.18, n.2, p.231-239, 2000.

SHAPIRO, S. S.; WILK, M. B. "An analysis of variance test for normality (complete samples)". Biometrika 52 (3-4): 591-611. doi:10.1093/biomet/52.3-4.591. JSTOR 2333709. MR 205384. p. 593, 1965.

SOUZA, R.F.; MADEIRA, N.R.; FIGUEIREDO, C.C. Soil, water and nutrient losses in an area cultivated with vegetables under a no-till system. **Revista Científica**, n. 1, p. 38-50, 2014.

STREK, H.J.; WEBER, J.B. Adsorption, mobility and activity comparisons between alachlor (Lasso) and metolachlor (Dual). In: SOUTHERN WEED SCIENCE SOCIETY, 35, 1982, Georgia. **Proceedings...p** .332-8.

THEISEN, G.; VIDAL, R.A.; FLECK, N.G. Reduction of *Brachiaria plantaginea* infestation in soya beans by covering the soil with black oat straw. **Pesquisa Agropecuária Brasileira**, Brasília, v. 35, n. 4, p. 754-756, 2000.

THELAN, K.D.; KELLS, J.J.; PENNER, D. Rotational crop response and volatilisation with FMC-57020. In: NORTH CENTRAL WEED CONTROL CONFERENCE, 41, 1986, Milwaukee.

Proceedings...p.48.

TIMOSSI, P.C. **Management of cover crops and integrated weed control in no-till soya**. 2005. Thesis (Doctorate) - Faculty of Agricultural and Veterinary Sciences, Paulista State University, Jaboticabal, 2005.

VELDKAMP, J. F. A revision of *Digitaria* Haller (Gramineae) in Malaysia. **Blumea**, Leiden, v. 21, n. 1, p. 1-80, 1973.

VIEIRA, R.C.B; BAYER, C.; FONTOURA, S.M.V.; ANGHINONI, I.; ERNANI, P.R.; MORAES, R.P. Liming criteria and critical levels of phosphorus and potassium in Latosols under no-tillage in south-central Paraná. **Revista Brasileira de Ciências do Solo**, v.37, p.188-198, 2012.

YODER, R. N. et al. Aerobic metabolism of diclosulam on U.S. and South American soils. **Journal of Agricultural and Food Chemistry**, Easton, v. 48, n. 1, p.4335-4340, 2000.

MIX
Papier aus verantwortungsvollen Quellen
Paper from responsible sources
FSC® C105338

Printed by Books on Demand GmbH, Norderstedt / Germany